CLIMATE CHANGE

© Copyright 2018 - All rights reserved.

This document is geared towards providing exact and reliable information in regards to the topic and issue covered. The publication is sold with the idea that the publisher is not required to render accounting, officially permitted, or otherwise, qualified services. If advice is necessary, legal or professional, a practiced individual in the profession should be ordered.
- From a Declaration of Principles which was accepted and approved equally by a Committee of the American Bar Association and a Committee of Publishers and Associations.

In no way is it legal to reproduce, duplicate, or transmit any part of this document in either electronic means or in printed format. Recording of this publication is strictly prohibited and any storage of this document is not allowed unless with written permission from the publisher. All rights reserved.

The information provided herein is stated to be truthful and consistent, in that any liability, in terms of inattention or otherwise, by any usage or abuse of any policies, processes, or directions contained within is the solitary and utter responsibility of the recipient reader. Under no circumstances will any legal responsibility or blame be held against the publisher for any reparation, damages, or monetary loss due to the information herein, either directly or indirectly.

Respective authors own all copyrights not held by the publisher.

The information herein is offered for informational purposes solely, and is universal as so. The presentation of the information is without contract or any type of guarantee assurance.

The trademarks that are used are without any consent, and the publication of the trademark is without permission or backing by the trademark owner. All trademarks and brands within this book are for clarifying purposes only and are the owned by the owners themselves, not affiliated with this document.

Table of Contents

- INTRODUCTION 5
- UNDERSTANDING CLIMATE CHANGE 11
- WHAT CAUSES CLIMATE CHANGE 16
- HOW CLIMATE CHANGE CAN AFFECT YOUR LIFE 20
- THE CONTROVERSY BETWEEN 23
- CLIMAYTE CHANGE AND GLOBAL WARMING 23
- ENVIRONMENTAL ISSUES AND CLIMATE CHANGE POLICY 29
- CONSEQUENCES OF CLIMATE CHANGE 34
- CHALLENGES OF CLIMATE CHANGE AND BIO-ENERGY 37
- KEY ELEMENTS OF CORPARATE CLIMATE CHANGE STRATEGY 48
- CLIMATE CHANGE ADAPTATION AND AGRICULTURE 53
- HOW TO PREVENT CLIMATE CHANGE 59
- GLOBAL WARMING TO CLIMATE CHANGE .. 63
- GETTING REAL ABOUT CLIMATE CHANGE .. 67
- CONCLUSION 71

INTRODUCTION

As is the case with many ecological matters nowadays, climate change is a hot topic for debate. Some will argue that the global climate is changing as a matter of natural course while others will point an accusing finger at humankind as the culprit leading to Earth's rising atmospheric temperature.

Whichever argument you side with, there can be no debate that the symptoms of a slowly rising fever have begun to show, especially in recent decades.

Whether these climbing temperatures are a result of ordinary natural mechanisms or of human irresponsibility does not need to be debated; we need to look at what climate change is, how quickly it is advancing, and what we can do to alleviate the problem.

Climate change is often used interchangeably with global warming. This is not completely accurate, however, because climate change refers to a change in average statistical data over prolonged periods of time regardless of what caused it.

Global warming is usually reserved to acknowledge climate changes brought on directly or indirectly by human activities. Again, we are not here to debate the causes but to recognize certain aspects of of the issue.

When all of us here on Earth are better versed in the machinations behind our gradually rising average temperatures, we can become more conscious of our contributions to climate change and more effectively alter our habits to lessen our own impact. For example, solar output is one possible explanation for the increased temperatures that lead to climate change.

The sun emits much more radiation today than it did 3 to 4 billion years ago. The sun is cyclical in the amount of radiation it gives off, leading to slight changes in climate every so often. An one hypothesis as to the varying amount of radiation that reaches the earth is changing concentrations of greenhouse gases in the atmosphere.

Billions of years ago, there was a larger concentration of these gases, allowing less radiation through, but trapping what did pass into the lower atmosphere. Temperatures rose and fell depending on the quantity of solar radiation remaining in the atmosphere.

Gradually, these greenhouse gases lessened in concentration and a sort of balance was achieved. Today, for whatever reasons, the gases have increased in accumulation again, leading several scientific bodies to conclude that they are playing a role in climate change.

It doesn't matter if humans are responsible or not, we can make the choice to do our part by not exacerbating the problem. One large change we can make is to seek alternative methods for powering our homes and vehicles.

Solar cells, fuel cells, wind power, and wave power are just some of the renewable energy sources we can adopt to limit our release of greenhouse gases into the atmosphere.

Think about it. If the sun is simply experiencing another of its natural cycles and raising Earth's average temperature, then that is just the way it goes. But why should we accelerate the process if we don't have to?

Imagine the amount of carbon-based exhaust that would be eliminated if homes didn't rely on electricity sources that spewed out fumes as a result of spinning turbines powered by petroleum fuels.

Vehicles equipped with fuel cells that simply give off water as a by-product add absolutely no greenhouse gases to the atmosphere.

There is still a lack of clear cut evidence that humans are the sole contributors to Earth's climate change, but we can erase that possibility altogether by changing the way our power needs are met.

Another possible explanation for climate change is the increased amount of aerosols in the atmosphere. Aerosols are tiny solid or liquid particles that remain in the atmosphere. It is not completely understood how much of an impact they currently have on climate change, but it is known that aerosols absorb and concentrate solar radiation.

Volcanoes, vegetation, sea spray, and forest fires are all known to contribute aerosols to the atmosphere. Human creations such as vehicle exhaust and industrial plants do as well. These anthropogenic aerosols account for about 10% of those currently found in the atmosphere.

Once again, there is no way to accurately measure just how much of an impact anthropogenic aerosols have on climate change, but there is no reason to continue to scatter particles that are known to have a greenhouse effect into the air.

Here is yet another case in which using the renewable energy of the sun, wind, and water would greatly reduce the amount of stress we place on the atmosphere to sustain temperatures that allow for life.

Regardless of what is leading to climate change, it is happening. Is it the ordinary path of life on Earth or is humanity burdening the atmosphere with elements detrimental to its function?

Whatever the answer, we can do our part to combat climate change by changing our attitudes and habits and adopt renewable energy as the first step in getting back on track.

This comprehensive guide to climatic change provide all the relevant facts and methods to overcome these natural problems.

Happy Reading.

CHAPTER 1

UNDERSTANDING CLIMATE CHANGE

Climate change can be defined as a change in climate variables, especially temperature and rainfall that occur gradually in a long period of time between 50 to 100 years.

Besides it should be understood that the changes caused by human activity (anthropogenic), especially those related to fossil fuel consumption and over-land.

So the changes caused by natural factors, such as additional aerosols from volcanic eruptions, are not accounted for in terms of climate change.

So the natural phenomena that lead to extreme climatic conditions such as cyclones can occur in a year (inter-annual) and El-Nino and La Nina, which can happen in ten years (inter-decade) cannot be classified into global climate change.

Human activity in question is the activities that have led to an increase in atmospheric concentrations, especially in the form of carbon dioxide (CO_2), methane (CH_4) and nitrous oxide (N_2O).

The gases then determines the temperature increase because it is like glass, which can forward to short-wave radiation which is not hot, but hold long-wave radiation that is heat. As a result the earth's atmosphere heats up.

The impact of climate change:

Agricultural sector will be affected through the reduction of food productivity caused by the increase in cereal sterility, reduction in area can be irrigated and decrease the effectiveness of nutrient absorption and spread of pests and diseases.

In some places in the developed (high latitudes) increase in CO_2 concentration will increase productivity because of increased assimilation, but in the tropics, that most developing countries, an increase of assimilation was not significant compared with respiration which is also increased.

On the whole, if adaptation is not done, the world will experience a decline in food production to 7 percent.

However, with continued levels of adaptation, meaning high costs, food production can be stabilized. In other words, stabilization of food production on climate change will cost very high, for example by improving irrigation facilities, provision of inputs (seeds, fertilizers, insecticides / pesticides) added.

In Indonesia, the scenario of CO_2 concentrations double from current rice plant production will increase to 2.3 percent if irrigation can be maintained. But if the irrigation system did not experience improvements in rice plant production will decline to 4.4 percent.

Warmer temperatures will cause a shift in vegetation species and ecosystems. Mountain areas will lose many species of original vegetation and replaced by lowland vegetation species.

Along with that the condition of water resources from the mountains will also be susceptible to interference. Furthermore the stability of land in mountainous areas is also disrupted and hard to keep the original vegetation.

This impact is not so apparent in the low latitude areas or low elevation area. If more and more forest fires are common in Indonesia, it was difficult to connect the incident with climate change, because most (if not all) incidents of forest fires caused by human activities associated with land clearing.

That happened in conjunction with the El-Nino events because of this phenomenon provides dry weather conditions that facilitate the occurrence of fire.

However, as described above El-Nino is phenomenon of nature which associated with extreme climate events in climate variability, not climate change in the meaning as described above. Increasing of population makes pressure on water supply, especially in urban areas.

At this moment there are lots of urban residents who have difficulty getting clean water, especially those who are low- income and low-educated or unskilled.

The impact of climate change which causes changes in temperature and rainfall will have an impact on the availability of water from the surface runoff, groundwater, and other reservoir shapes. In the year 2080, there will be 2 - 3.5 billion people will experience water shortages.

CHAPTER 2

WHAT CAUSES CLIMATE CHANGE

Climate change has a number of of causes, both natural and manmade. However, the last two hundred years have seen changes to the climate occurring more rapidly than ever before.

This is predominantly attributed to global warming which, in turn, has been caused by the burning of fossil fuels since the industrial revolution.

The resulting release of greenhouse gases and particulate matter also cause a range of associated processes that also contribute to climate change.

Scientists have used data from ice cores, and other sources to record the planet's climate for the last 800,000 years. This record shows the fluctuations of temperature, rainfall and sea level that have occurred in this time.

This cycle of rise and fall is a result of a complex system of events including solar activity, ocean currents, polar ice caps and atmospheric pressure gradients.

Events such as meteor strikes and volcanic eruptions also have significant impacts on the global climate and have been known to cause the onset of an ice age. While changes to Earth's climate have always occurred, the frequency of the change in the last two hundred years is alarming scientists and governments around the world.

Global warming has been linked to the rapid change in climate observed since the industrial revolution. The burning of fossil fuels releases greenhouse gases that in turn trap more of the sun's heat in the planet's atmosphere. This raises global temperatures causing a range of other impacts that can affect the climate.

One of the most serious impacts of global warming is the melting of ice caps in the north and south poles.

They are both important in regulating the planet's climate by maintaining polar albedo (the reflection of solar energy back to space) and the ocean currents that affect major weather systems.

As the ice melts less heat is reflected, and the planet becomes warmer. Fresh water entering the ocean in large volumes can also alter ocean currents, further affecting weather systems resulting in changes to the planet's climate.

Changes to the surface of the planet's land masses have also been found to affect climate. Deforestation, urban development and agricultural practices all change the amount of the sun's heat reflected or absorbed by the surface. The carbon cycle is also disrupted; resulting in less atmospheric carbon being sequestered, further increasing global warming that leads to climate change.

The burning of fossil fuels releases particulates into the atmosphere causing acid rain. The particles in the atmosphere can also reduce global warming by reflecting sunlight before it reaches the planet's surface.

This process is known as global dimming and has actually been found to counteract the causes of climate change in some instances.

Scientists have even suggested purposely releasing sulphur particles into the stratosphere to increase the amount of sunlight reflected back to space.

While this would slow climate change, there are a number of negative impacts, particularly on the health of living creatures from the higher levels of particles in the air that they breathe.

Although the planet's climate is influenced by a complex system of natural processes, global warming is the principal cause of recent climate change.

Reducing our reliance on fossil fuels, therefore cutting greenhouse gas production is the most important action that humans can take to minimize climate change.

CHAPTER 3

HOW CLIMATE CHANGE CAN AFFECT YOUR LIFE

Climate change has a very profound effect on our day to day life. But off late we are seeing instances where our weather patterns have changed and there is an increase in erratic climatic changes all over the world. The sudden climatic change which is occurring is mainly due to global warming.

All over the world there are many countries that have a fast and developing the economy and as the countries concentrate on their growing economy our planet is being left in a very bad shape.

The people have neglected the environment running behind money and the climatic changes have come back to haunt the people.

People were able to live comfortable lives as they were adjusted to the climate and environment, but our own activities have caused the climatic changes which are mostly due to global warming which has left us uncomfortable.

The temperature during the summer is on the rise and some places are experiencing very harsh winters.
The effects of climatic changes are affecting people, wildlife and our environment. In basic terms, Global warming is the increase in temperature of the earth's surface and atmosphere.
The major reasons of climatic change are the burning of fossil fuel, the greenhouse gases released in the emissions of houses, factories, cars etc.
The climatic changes have been the reasons of melting of glaciers in the Polar Regions, occurrences of large number of hurricanes, tsunamis, high variations in climate, change in the rainfall distribution etc. due to the increase in temperature there have been formation of high and low pressure regions which have caused the formation of hurricanes and extreme climatic conditions.
Man, by his carelessness has led to the climatic change thereby causing destruction to our own property.
Hurricanes and other changes have been devastating to both man and animals. Due to the sudden change, it has led to destruction of property, loss of many lives, occurrence of diseases, floods etc.
Floods which are caused due to the climatic change can cause stagnation of water which is

perfect breeding grounds for many vectors like mosquitoes which causes diseases like malaria, filarial etc.

Another effect of global warming has been that due to depletion of ozone layer, there is no barrier to prevent the harmful rays of the sun. Thus the temperature increases and also chances of skin cancer due to ultra violet rays.

These rays may also create mutations which will be transmitted to next generations and can cause genetic disorders.

The melting of ice has led to slow increase in the sea level and if this continues, it will lead to complete submerging of many islands under the sea in the future years to come.

The harsh effects of the climatic changes have made people to think about the reasons for the global warming.

It has caused an increase in the effort by people to reduce global warming by switching to different energy source other than fossil fuels, recycling, planting trees, electricity conservation and awareness programmes.

Even, the governments, are signing treaties and under taking projects to reduce global warming and thus prevent the climatic changes.

If we want to ensure that there are no more adverse climatic changes then we need to take steps to prevent global warming so as to ensure that earth is a place to live for the next generation.

CHAPTER 4

THE CONTROVERSY BETWEEN CLIMATE CHANGE AND GLOBAL WARMING

The term climate change and global warming are often used interchangeably as they refer to the same environmental problem. Some people prefer to use climate change as climate is more evident to us and climate change is not as controversial as global warming.

It is probably okay to use a either term but for those who like precision in language, we will take a tour through climate science to sort out the difference.

Climate and Weather: It is said that no one can predict the weather and that is true as no one can predict very far in advance whether it will rain or storm or how cold or hot it will be. However, if we observe the weather of a region over a long period of time, a pattern emerges.

That pattern is the climate and, though we cannot predict the weather accurately, we have a much better chance of predicting climate.

If we observe such things as the high and low temperatures, the amount of rain, when the first frost and the last freeze usually occurs, a pattern emerges.

The climate is ⬜uite important to us as it determines the crops we grow, the types of house we build, and the clothing we wear. Climate determines the plants, animals, and insects that live in our region and even the types of health problems and diseases.

The factors that determine climate have been observed to vary slowly with time and we expect the climate in a region to remain relatively stable over long periods.

Climate Change: In the last century, and particularly in the last three decades, we have noticed that climates in many regions of the Earth are changing.

The daytime high temperatures are higher, the nighttime lows are warmer, the patterns of drought and rainfall have changed, and storms seemed to have become stronger. Frost occurs later in the year and the last freeze occurs earlier, which has caused gardening zones to move.

The ranges of many species of plants, animals, insects and bacteria have shifted, and there has been invasions of non-native, sometimes invasive, species into new areas. Our observations have shown that the climate is definitely changing, and those changes are sure to have consequences for us.

Global Warming: Since the early 1800's, scientists have been concerned with whether our use of fossil fuels has affected the temperature of the Earth. With an increasing understanding of the role greenhouse gases play in stabilizing the temperature of the Earth, scientist wondered whether burning fossil fuels might affect the energy balance of the Earth.

Burning carbon fuels releases carbon dioxide, CO_2, which they knew to be an important greenhouse gas and there was speculation about whether an increase of CO_2 in the air could actually cause the Earth to warm.

Critics of the idea argued that water was a much more important, that the relatively small amount of CO2 in the air would not make a difference, and that the amount of CO2 man produced was minuscule compared to what was already there.

The Role of CO2: With a better understanding of the atmosphere and the advent of computers, we are able to calculate the climate sensitivity of the Earth to CO2. By doubling the concentration of CO2 in the air would cause a 3 to 4 °C increase in the Earth's temperature.

A number of studies have confirmed that the concentration of CO2 in the air is small, it accounts for about 25% of the greenhouse effect. Certainly, increasing the amount of CO2 in the air should cause the Earth to warm.

In the last century, our emission of CO2 has increased from a minuscule amount to over 50 billion tons annually and the concentration of CO2 in the air has risen from 280 parts per million (ppm) to 385 ppm. But, has that caused global warming?

The Temperature Scorecard: The temperatures over the Earth vary widely from place to place with the weather and the season. However, the temperature of a particular place measured over a long period of time has a pat-tern and we can use the pattern as a scorecard.

There are temperature records that go back to about 1850 and these have given us a way to keep track of whether the Earth is warming. By using ships logs, weather stations, and satellite measurements, NASA has compiled the Earth's annual mean temperature from 1880 to the present.

Though it varies widely from year to year, the Earth's annual mean temperature shows an upward trend and the Earth is in fact getting warmer.

The scorecard shows that over the last century the Earth has warmed about 1.3°F, which does not sound like much. However, since that is the average over the whole Earth, it represents a tremendous amount of energy and it is the energy in the atmosphere that drives our weather and determines our climate.

So there we have it, a cause and effect relationship. Climate change is caused by global warming, which in turn is caused by the increasing CO2 in the atmosphere, and the CO2 is increasing because of our use of fossil fuels.

Though it is probably irrelevant whether we call it climate change or global warming, it is very relevant that we understand the relationships and think about our role. The way we use fossil fuels has consequences for us and for the rest of the species on the planet.

CHAPTER 5

ENVIRONMENTAL ISSUES AND CLIMATE CHANGE POLICY

Climate change is inevitable. It has, and always will be, a feature of our planet. Why then are we so worried about it? We worry because in our modern world of growing human numbers and affluence, rapid climate change affects us directly.

Changes to rainfall, temperature, frequency and intensity of severe weather, shifts in seasonality, and other locally significant effects, such as seal level rise and melting glaciers, are the obvious consequences of climate change.

These effects compromise food security, our water supplies, economic stability, and in extreme situations threaten lives.

Somewhere in our subconscious, we are surely aware that climate change effects are more acute than they used to be. A world containing 7 billion people who with the help of their immediate ancestors, have modified every corner, is not as buffered as it was.

Over time we have modified the environment to feed, clothe and shelter the generations. We have cut down trees, ploughed fields, diverted rivers and reared livestock. The ability to make such modifications, and the responsiveness of the environment to the changes we have made, is why there are so many of us. These modifications to habitats have compromised environmental performance.

Recall that many a conservation scientist has warned of the dangers of biodiversity loss. They say that loss of diversity means fewer options for adaptation and delivery of ecosystem services. Where habitats are changed biodiversity is lost and nature is not as robust and resilient as she used to be.

Consider a forest cleared for a wheat crop.

Wheat is an annual grass that dies back once the seed heads have matured, so part of the year there is only straw stubble in the field. Commonly farmers will plough in or even burn this stubble to leave the soil bare for many months. Exposed soil loses moisture, carbon and it's biological activity.

Dry, exposed soil is vulnerable to the wind and is readily eroded under heavy rain. Each year the grain crop feeds us only over time soil structure, moisture retention, and biological activity decline. Unless we apply fertilizer and insecticides yields decline too.

This bare soil and single species crop system that becomes dependent on inputs is not resilient to climate change. Warmer and drier or colder and wetter, extreme events and changed seasonality all affect productivity.

The original forest is well buffered against these effects. Trees are long-lived with deep root systems. Tree canopies and a layer if leaf mulch protects the soil surface to help retain moisture and maintain biological activity in the soil. Shifts on weather have little overall effect. Unfortunately, it is not possible to make bread from trees.

Whilst floods and drought deliver the sound bites and photo opportunities for climate change, intuitively we know that the modified landscapes that provide us with food and water are vulnerable to climate shifts. It is a worry. Not surprisingly we expect our leaders to implement policies some action to alleviate our concerns.

Humans are a action orientated species. The population wants to see something done. The crux of the vexing debate over climate change policy is that something can be done about these changes to the climate. Alternatively, nothing can or needs to be done, depending on your point of view.

It also assumes that policy will not only generate that 'something' but that what is done will ultimately fix the problem. It may be worth a moment away from rhetoric and spin to consider these assumptions once again.

The current policy debate is about greenhouse gas emissions. The premise is that human activities in the last 200 years in clearing land for agriculture and livestock, and in burning fossil fuels for energy and transport have triggered warming through an increase in the concentration of greenhouse gases in the atmosphere. This we know, almost to the point of dogma. We also knows that decreasing emissions is the chosen policy solution.

And so the political debate has become how to reduce emissions. What policies will slow energy consumption and the emission intensive activities without damaging economic activity?

Is it a direct tax on emissions, a market trading scheme for emission credits, subsidies for alternative energy generation, regulation to limit emissions from vehicles, or combinations of a host of other options that are available.

The debate has rarely covered the consequences of climate change. It has focused on action being taken that will fix the problem - actions to stop climate change.

Future historians will applaud actions to shift from fossil fuel to dependence. However, they will be totally confused by such a single focus. "Why," they will say, "was so little done to change land management when the conseduences of climate change for food production and water supplies were so obvious."

This is the real environmental issue of climate change policy. It is not a singular problem at all.

CHAPTER 6

CONSEQUENCES OF CLIMATE CHANGE

Climate models are computer operated simulation programs that give climatologists a way to predict climate changes due to the changes in natural concentration of greenhouse gases and alterations in earth's natural ecosystems and habitats.

Climate scientists have been collecting data about the climate and temperature changes from around the world for a long period of time and when this data is fed into the climate model, future climate changes and its consequences can be predicted to a certain extent.

Although the consequences of global climate change may seem to be the stuff of Hollywood - some imagined, dystopian future, the melting ice of the Arctic, the spreading deserts of Africa, and the swamping of low lying lands are all too real.

We already live in a 'age of consequences', one that will increasingly be defined by the intersection of climate change and the security of nations.

AMERICAN GEOPHYSICAL UNION

Climate Change Predictions:

Following are the results of climate models predicting various consequences that may occur due to climate change:

The rise of Global Average Temperature: The global average temperature is expected to rise by 0.64-0.69 degree Celsius by the years 2011-2030 compared with the years 1980-1999.

The predictions show a gradual rising trend of earth's average atmospheric temperature and will force several biomes to shift to high altitude areas or cooler regions. Areas that will be most affected are Alaska and the Arctic regions.

Disruption of Ocean circulation: It is very likely that the climate change will bring in disruption of ocean circulation patterns that keep northern European countries in a suitable temperate range.

Melting of Ice Sheets: With a global temperature rise in the range of 1-4° Celsius, scientists are predicting partial melting of the Greenland and the West Antarctic ice sheet.

This prediction is however accepted by IPCC (Intergovernmental Panel on Climate

Change) under a level of uncertainty because global temperature rise of 1-4 degree Celsius may take hundred's or even thousands of years to occur.

Extreme Weather Events: Due to global temperature rise, intensity of natural events like heat waves, precipitation and evaporation will increase and it will cause extreme weather events such as severe droughts, heavy rainfall, hurricanes, tornadoes etc. These severe droughts will propagate through many critical ecosystems in North America, southern Europe and Asia, northern Africa, Canada, and Alaska and will result in increasing dry-land areas.

Not only these extreme weather events will cause loss of life and property, it will also increase stress on available natural resources to meet the demand of lost ecosystems.

CHAPTER 7

CHALLENGES OF CLIMATE CHANGE AND BIO-ENERGY

Scientific and technological advancements have revolutionized the entire human civilization in a truest sense. It has brought us to a point where we can assume that everything we imagine and conceive is practically achievable.

Nowadays, when our lives are surrounded by so much of digitalization and hi-tech machinery, when the rapidness of development and research is so impressive, it is fairly easy to forget the inescapable fact that we are damaging our mother world at an unprecedented pace.

So often in course to satisfy our hunger of attaining economic supremacy and industrial feasibility, we fail to realize that we are actually deteriorating our natural resources.

We, along with all our advancements are disturbing the ecological and environmental balance at such a frantic pace that the entire human history has never witnessed before.

And while doing this, we have provoked the nature's need for revenge. We have made ourselves more vulnerable to stern temperatures, floods, hurricanes, typhoons, droughts, excessive rainfall, and now it is a critical time to understand that if we continue to exploit nature and affect climatic balance and do nothing to alleviate this issue, we are bound to face devastating consequences.

Climate changes pose clear, catastrophic threats. We may not agree on the extent, but we certainly can't afford the risk of inaction. To better understand the issue, we must first study what are climate changes and which factors are responsible for them. The term climate change is often used interchangeably with the term global warming.

The phrase 'climate change' is growing in preferred use to 'global warming' because it helps convey that there are [other] changes in addition to rising temperatures."

Climate change refers to any significant change in measures of climate (such as temperature, precipitation, or wind) lasting for an extended period (decades or longer).

Global warming is an average increase in the temperature of the atmosphere near the Earth's surface and in the troposphere, which can contribute to changes in global climate patterns.

Earth maintains its average temperature by a natural and self-automated warming system of gases which surround it.

Carbon dioxide and other gases like methane, Nitrogen dioxide and Chloro Flouro Carbon (CFC) keep the earth warm by trapping solar heat in the atmosphere. This trapped heat is crucial in keeping earth's temperature within a range where it is habitable.

However, the uncontrollable increase in the emission of Carbon dioxide and other warming gases over the decades has thickened these atmospheric boundaries which are now retaining much more heat than the acceptable range.

Further, the increase of carbon dioxide and other gases in the atmosphere has also enhanced the "Greenhouse Effect" in which more heat is generated. This excessive amount of heat has disarrayed earth's natural thermo-equilibrium resulting in the form of global warming with all its associated climatic effects.

The history of the planet has been characterized by frequent changes in climate. Apparently, climate change is a natural phenomenon occurring since several thousand years.

Environmental scientists insist that earth's temperature has always been on a gradual rise with no or very limited impact on the environment on whole.

This gradual trend spanning over a period of 650,000 years shows a gradual rise which scientists initially thought of as a "slow motion catastrophe" an unexpected to show its earliest consequences generations later.

Needless to say, time has proved this estimations erroneous since signs of the climatic changes due to increased earth temperature have accelerated alarmingly in last two centuries.

The graphical relation between time and earth's temperature proves a dramatic and unparalleled shift in the trend with temperatures increasing many times faster than ever in the recorded history.

Based on data from the UN's Intergovernmental Panel on Climate Change, it is estimated that the mean global surface temperature has increased by about 0.3 to 0.6 degree Celsius since the late 19th century to the present, and an increase of 0.2 to 0.3 degree over the last 40 years.

This increase is likely to have been the largest of any century during the past 1,000 years. The current rate of increase of greenhouse gases is unprecedented during at least the past 20,000 years.

And with the help climatic models based on mathematical simulations, it is predicted that by the year 2050, global temperature would be rose around 5 degrees Celsius with some severe and unavoidable impacts.

There are a number of natural factors responsible for climate change. Some of the prominent ones includes continental drift, volcanoes, ocean currents, the earth's tilt, and comets and meteorites.

But the Anthropogenic Factors is the real culprits which have induced such an uncontrollable emission of carbon dioxide and other gases and therefore elevated average temperatures.

Anthropogenic factors are human activities that change the environment and influence climate. In some cases, however, the chain of causality is clear and unambiguous while in others it is less clear.

Various assumptions for human-influenced climate change have been debated over the years but it is only now widely accepted without any doubt that the major cause of climate change are the human activities.

Even those who up until as recently as a few years ago were not convinced that humans have an impact on the climate, now admit that scientific evidence exists that this is happening.

The Industrial Revolution, starting at the end of the 19th Century, has had a huge effect on climate. The invention of the motor engine and the increased burning of fossil fuels in form of coal, oil and natural gas have increased the amount of carbon dioxide in the atmosphere. Since then, the human consumption of fossil fuels has elevated CO_2 levels from a concentration of ~280 ppm to ~387 ppm today.

These increasing concentrations are projected to reach a range of 535 to 983 ppm by the end of the 21st century. It is now known that carbon dioxide levels are substantially higher now than at any time in the last 750,000 years. With the prevailing concept of global economy and the accelerated industrialization of developing countries like India and China, 70 million tons of CO_2 is dumped into atmosphere everyday.

In addition to CO_2, Methane is another important greenhouse gas in the atmosphere. About ¼ of all methane emissions are said to come from domesticated animals such as dairy cows, goats, pigs, buffaloes, camels, horses, and sheep.

These animals produce methane during the cud-chewing process. Methane is also released from rice or paddy fields that are flooded during the sowing and maturing periods.

When soil is covered with water it becomes anaerobic or lacking in oxygen. Under such conditions, methane-producing bacteria and other organisms decompose organic matter in the soil to form methane.

Nearly 90% of the paddy-growing area in the world is found in Asia, as rice, is the staple food there. China and India, between them, have 80-90% of the world's rice-growing areas.

Methane is also emitted from landfills and other waste dumps. If the waste is put into an incinerator changes triggered by such gases are anticipated to cause an increase of 1.4-5.6 °C between 1990 and 2100.

The cement manufacturing industry in particular, contributes CO_2 when calcium carbonate is heated, producing lime and carbon dioxide, and also as a result of burning fossil fuels. The cement industry produces 5% of global man-made CO_2 emissions, of which 50% is from the chemical process, and 40% from burning fuel.

The amount of CO_2 emitted by the cement industry is nearly 900 kg of CO_2 for every 1000 kg of cement produced. [out] or burnt in the open, carbon dioxide is emitted.

Methane is also emitted during the process of oil drilling, coal mining and also from leaking gas pipelines (due to accidents and poor maintenance of sites).

A large amount of nitrous oxide emission has been attributed to fertilizer application. Another gas, nitrous oxide, emitted in a very large from fertilizers can cause serious damages. These climate

One of the other major factors of climate change is Increased Land Use. Agriculture practices, irrigation and deforestation are fundamentally changing the environment.

Due to increased urbanization and industrial growth, forests are being cut down which act as "Carbon sinks". As a result, that the extra carbon dioxide produced cannot be changed into oxygen.

Accepting the factors that are causing it, an overwhelming majority of scientists today agree that climate change is real and poses very serious global threats. These climate changes have already shown some shocking and horrific signs around the world.

They are by now affecting lives of millions of people throughout the world and are expected to get far more ruthless in future. In particular, many developing countries though they have contributed to the least in the process of climate change will be the ones at the greatest risks to face the consequences.

Burning biomass efficiently results in little or no net emission of carbon dioxide to the atmosphere, since the bio-energy crop plants actually took up an equal amount of carbon dioxide from the air when they grew.

However, burning conventional fossil fuels such as gasoline, oil, coal or natural gas results in a increase in carbon dioxide in the atmosphere, the major gas which is thought to be responsible for global climate change.

Some Nitrogen Oxides inevitably result from biomass burning (as with all combustion processes) but these are comparable to emissions from natural wildfires, and generally lower than those from burning fossil fuels.

Other gas emissions are associated with the use of fossil fuels by farm equipment, and with the application of inorganic fertilizers to the bio-energy crop. However, this may be offset by the increase in carbon storage in soil organic matter compared with conventional crops.

Utilization of biomass residues which would have otherwise been dumped in landfills (e.g. urban and industrial residues) greatly reduces greenhouse gas emissions by preventing the formation of methane.

In addition, bio-energy can effectively be used in almost every industrial, manufacturing and home application throughout the globe.

Wood, construction waste, landfill gas, and liquid bio-fuels like bio-diesel and bio-oil can be used to produce energy that can be converted into electricity and heat. Liquid bio-fuels like ethanol, bio-diesel, and bio-oil can be used to power cars and other transportation.

Being the fourth largest resource of energy after coal, oil and natural gas, the energy produced from the bio-mass can fulfill up to 14% of the world's total primary energy demands and recent statistics show that only 10-15% of the total potential bio-energy sources have been used so far by the human population worldwide.

Along with its remarkable and efficient outcomes in decreasing the world's carbon emission and fulfilling a considerable portion of the global demand for energy, Bio-Energy from the bio-mass also has several major socio-economical benefits.

These benefits include increased rural income and reduced levels of poverty in developing countries, restoration of unproductive and degraded lands and promotion of economic development, diversifications of agricultural outputs, reduction of energy dependence and diversification of domestic energy supply, increased investments in land rehabilitation and effective usage of waste products.

A recent economic survey found out that bio-energy creates more permanent jobs than any other energy sources with a decrease in unemployment and increase in per capita income which contributes to a much healthy life style. It can also be instrumental in reducing food prices and ensuring food security throughout the world.

In keeping a watchful eye over the huge opportunities the usage of bio-energy can offer, every possible step should be taken by the

United Nations and the state governments all over the world to replace fossil fuels with bio-fuels. Since it is practically unrealistic to completely replace fossil fuels, intense attempt should be made to utilize as much of the natural energy resources as it is possible.

On individual level too, we should adapt to these climatic changes and change our live styles in order to bring the total carbon emission under control.

Driving less, driving a fuel-efficient car, preferring gas over oil, saving electricity, using lesser papers and planting more trees can be some of the small choices each human can makes to save the earth from rapid destructions of the climate change.

It is not only a environmental issue. It is inherently linked with our lives on political, social, economic, ethical and more than anything else, on moral grounds.

We do not as a society lack in resources and capabilities but it is a high time that we confront the challenges of the climate change with utmost determination and a collective strategy.

CHAPTER 8

KEY ELEMENTS OF CORPORATE CLIMATE CHANGE STRATEGY

The greatest environmental challenge of our time is without a doubt climate change. Over the coming years and decades this will have a stark influence on our lives.

Not only through the physical changes in the climate, but also through changes in our energy consumption, travel patterns and many other aspects of our life. Companies will not be unaffected by these changes.

Markets will change, as will client requirements and there will be a steep rise in legislation with regards to climate change and greenhouse gas emissions.

Developing a comprehensive corporate climate change strategy is therefore a essential part of good business management. This chapter gives a brief description of the elements that such a strategy should include.

The management of climate change should adopt two distinct strategies: mitigation and adaptation. Mitigation deals with the reduction in greenhouse gases to the earth's atmosphere.

Carbon foot-printing and carbon accounting form typical measures that are used within the framework of this strategy.

Adaptation revolves around measures that address the changes and vulnerabilities of the organization that will be present as a result of the inevitable physical change in our climate. A good corporate climate change strategy addresses both mitigation and adaptation.

1. Understanding greenhouse gas emissions of the organization

An understanding of the organization's greenhouse gas emissions is fundamental in every credible corporate climate change strategy.

To achieve this a comprehensive greenhouse gas inventory should be made. The inventory is a first step to manage and reduce the emissions, which are also key elements of the strategy.

The inventory combined with the management and reduction form the core of a greenhouse gas accounting system.

At this moment, there are two internationally recognized systems for greenhouse gas accounting available. These are the ISO 14064 standard and the Greenhouse Gas Protocol.

2. Understanding climate change vulnerability

Most large organizations have started to make a greenhouse gas inventory and have adopted measures to reduce their carbon footprint. As I described in the introduction, adaptation also forms an important part of a comprehensive corporate climate change strategy.

To develop an understanding of an organization's vulnerability to changes in the climate a review of the operations, products and services, transport needs should be made as well as an understanding of the physical changes to the climate in which the organization operates.

There are no clear guidelines on how to develop an understanding of the vulnerability to changes in the climate, although the United Nations Panel on Climate Change has published a range of reports documenting the predicted changes to the climate in each of the global regions over the course of the current century.

In addition to these global publications, many governments have funded research predicting the potential change in climate at a national level.

3. Commit to reduction in greenhouse gas emissions

As part of credible strategy, a commitment to reduce greenhouse gas emissions should be made at the most senior level in the organization.

4. Develop internal corporate capacity

Development of internal capacity is an essential element to drive through improvements to both the greenhouse gas emissions and the climate change vulnerability.

It is the people within the organization that are best placed to identify practical solutions that will support any improvement programme.

Ensuring a adequate level of knowledge and understanding within the members of staff with regards to the climate change issues that are facing the organization is therefore a key element in the strategy.

5. Work with supply chain and engage with stakeholders

A significant contribution to the carbon footprint of a product that a organization produces is made during the processing by suppliers.

In addition an organization should be aware of the climate change vulnerability of its own supply chain. The strategy should therefore address the role of the supply chain in managing climate change issues.

6. Adopt and implement improvement action plan

The first five elements provide a thorough understanding of the issues facing an organization as a result of of changes to our climate. Based on this understanding an improvement plan should be drawn up.

As with any credible action plan the actions should consist of SMART targets, be fully funded, and be relevant to the organizations climate change threats and opportunities.

7. Publish an annual report

The final element is the publication of an annual report that demonstrate progress against the action plan, re-affirms the commitment to improvement and accounts for the carbon emissions.

CHAPTER 9

CLIMATE CHANGE ADAPTATION AND AGRICULTURE

Climate Change (CC) is the among the most important global environment concerns. CC is having significant impacts on the most vulnerable communities.

Considering characteristics, agriculture can be considered as one of the most affected sector and thus most vulnerable to climate change.

Mitigation and Adaptation:

There are several factors and sectors which are responsible for Green House Gas emissions and thus climate change. Agriculture sector remained most unmanaged till yet due to i) non-point source emission of GHGs, ii) need and priority of higher productions to feed the ever growing population. As mentioned above, agriculture sector is also most susceptible to climate change contributing into the vicious cycle.

Inefficient agricultural practices causes the GHG more emissions, which accelerating the CC, resulting into the lower production.

To compensate this, farmers tend to put more resources whenever and wherever are available in terms of fertilizer, water etc. which are the main factor for inefficiencies in the farm sector.

Effects of Climate Change and Solutions:

In Bundelkhand region, calculations of the climate reflect that after 25 years, rainfall is expected to reduce by 20% and the productivity of the existing crops may reduce by 15%. Decreasing landholding per farmer is going to have enhanced negative impact of these projections.

Thus the expected facts after 20-25 years would be:

Reduced rainfall, thus
Less availability of surface water
Deepened ground water level
Low returns of rain fed agriculture
Smaller landholdings
High input costs in agriculture
Lower profitability per unit of land
Higher risks of crop failures

Above estimations on reduced productions are based on the existing crops and practices. Our adaptation strategies need to be focus on the methods which can nullify or reduce the effects of the unreliable climatic conditions. Such as:

Adopting low water requiring/ drought resisting crops/crop verities

By reducing the water losses in agriculture and by irrigating the crops/plants not the land

Using improved practices to save the applied water like mulching etc.

Reducing the input costs by appropriate application of inputs

Adopting and promoting the alternative models for duel/multiple use of land like different forms of agroforestry or multi-cropping

By having fall back options like alternate income sources/crop insurance etc.

Where will there is a Way

Approach which has been adopted to take up and fight with climate change is two sided, i) Identifying the technical options and interventions required, ii) making community aware and helping them to take lead to fight climate change.

Identification of interventions

Drought resistant varieties are available at different stakeholders. However, the adoption is not much as farmers are not aware about these varieties and characteristics of these crops/crop varieties.

Such varieties have been promoted as they are the important option which can provide reliable production level even in extreme conditions.

There are various methods and ways available for reducing the water losses in irrigation. Sprinkler and drip methods can reduce the water application significantly.

Similarly, mulching can reduce the water loss from the soil.

For flood irrigation also contour cultivation, and making the check basins or borders depending on the land slope and soil type, can increase the water use efficiencies significantly.

Information on these options is being made available to the farmers through different stakeholders including the research organizations.

It has also been observed that the number of seeds we use per acre of land is quite higher than required for optimum production. This, not only increase the input costs but also reduce the productivity by creating unnecessary competition within the crop itself.

Appropriate use of manure and fertilizers, minimal tillage is the practices which reduces the input costs and enhance the profitability of the particular crops.

Agro-horticulture and other forms of Agro-forestry can be one way of reducing the dependability of the output of our efforts on the erratic monsoon, thus, stabilizing the incomes of the household and also its distribution across the year.

For the purpose of reducing the risks of crop failure, government is also promoting the efficient irrigation systems like drip and sprinkler for agro-horticulture use.

The government has also various other schemes to protect or at least protect partially ourselves from the risks of the crop failure from several reasons.

National Agriculture Research Scheme (NAIS) is one of them. We can get our crops insured under the scheme and in case of crop damage due to any of the enlisted reasons, we can get the benefits of the schemes.

It is also worth knowing at this point of time that quite of the options available are more accessible for the groups. Let's join hand together to access the information and take the challenges posed by climate change.

Creating Institutional Systems for Adaptation

Adaptation is the community and social initiative, which needs community actions. For sustainable farming society initiative, three tier institutional systems has been formed, at village level farmers groups consists of active farmers.

The objectives of the farmers clubs is to keep interactions and ensure access of farmers to knowledge and resources bases in terms of the meetings with scientists and extension workers, ensuring appropriate seed varieties and technologies etc. At second tier representatives of the farmers groups have formed the cluster of the farmers group.

These clusters are responsible for managing and creating cluster (cluster of geographically nearby villages). Top level tier is the federation level institution which takes the responsibility to resource building for all three levels of the institutions and establishing backward and forward linkages with markets.

It can be said that the approach for creating the institutions involving the target group and disseminating the appropriate practices and technologies is an appropriate response to climate change suiting both the adaptation and mitigation strategies.

CHAPTER 10

HOW TO PREVENT CLIMATE CHANGE

We probably need to accept that natural climate change is real, has been around ever since there was an atmosphere stable enough to create a climate and has causes that are definitely beyond us.

We will not be able to

Stop the earth from wobbling in its orbit,

Alter the sequence of sunspot activity and the subsequent energy pulses from the sun, or

Moderate volcanic eruptions and their gaseous emissions.

We are stuck with what these mechanisms bring. In one sense we are passengers on the roller coaster of global climate change and even to pretend we can be preventing climate change is just a little crazy.

But Google 'preventing climate change' and you will find plenty of websites claiming otherwise. There are lists of actions individuals can take for what can be done preventing climate change.

Here is a list from a green organization:

Change a light

Drive less

Recycle more and buy recycled

Check your tires

Use less hot water
Avoid products with a lot of packaging
Adjust your thermostat
Plant a tree
Turn off electronic devices when not in use
Stay informed
And here is another
Make your home more energy efficient
Make your lifestyle greener
Change your habits at work
Educate future generations
Get involved in the fight against climate change

The assumptions behind these and dozens of other similar action lists are familiar.

Climate change is all about global warming that, in turn, is all about greenhouse gas emissions and that is something we can stop.

Now before you cry heretic at your screen and click away to a greener site, please pause a moment.

Stop to think.

Even if everyone actually did all these things we would not prevent climate change. The climate system is huge and beyond our control.

This is not to say that our actions in burning fossil fuels and clearing land have not contributed to the next phase of change, they probably have. But even if it were possible, a simple reversal of the actions would not reverse the climate effect our actions have set in train.

Ask yourself if we really should enter into a war on climate change that we really can't win. Plus, we know that war is bad. Even so-called 'just wars' are bad.

Anything that we fight against we resist and the more we push the harder it is for us. We have wars on everything from terror to drugs and yet these things have not gone away.

It turns out that we should be proactive and carry out the actions on these lists to reduce greenhouse gas emissions (except maybe the fighting suggestion).

Not because we will be preventing climate change but because we need to transition away from economies based on fossil fuel because that energy source will soon become scarce and expensive.

We also need to be smarter about resource use now that there are unprecedented numbers of us and the frugality at the heart of all the suggested actions is a great way to start.

The risk in accepting prevention of climate change as plausible, let alone possible, is that we put our considerable energies in the wrong place.

If all we do is keep greenhouse gas emissions down then we will be in a mess when sea levels rise, droughts hit and the storms still come.

We should be wise and energize ourselves around adaptation, get closer in our understanding of how nature works to better know her limits, and equip ourselves for a world of 9 billion souls. Otherwise, like the Viking king we, will have to hang up our own gold crowns on a crucifix.

CHAPTER 11

GLOBAL WARMING TO CLIMATE CHANGE

The simplest explanation is that global warming is climate change in one direction. Specifically it is where net energy in the atmospheric system has increased sufficiently for a measurable warming. More energy in the system means that in many parts of the world it gets hotter.

One way that the Earth gets hotter is due to an increase in the atmosphere of gases that result in the greenhouse effect.

The brief explanation of the greenhouse effect is that the wavelength of energy arriving on the Earth's surface through the atmosphere is shorter than that reflected back towards space, a simple law of physics.

Some gases in the atmosphere let the shorter waves through but block the longer wavelengths, so the energy is bounced back down to Earth, trapped, effectively heating things up.

In the absence of the greenhouse effect mother Earth would be frigid and the planet's surface too cold for water to be liquid. The logic of this important phenomenon is that warming will happen when the concentrations of greenhouse gases in the atmosphere increase.

And this is where the confusion comes because we have attributed the current warming trend to a specific cause - us.

But climate can also cool. This is because there can be a reduction in the net energy balance. During periods of lower atmospheric energy in the geological past, the surface if the Earth has been much cooler than the present.

These periods were so cold that we called them ice ages, and there have been lots of them, some short, others long and intense.

So climate change means any change in climate conditions, warmer or colder. Global warming is a directional shift to a more energetic, hotter and more dynamic atmosphere. It is a phase of global warming that the scientific data suggest we are experiencing now.

This is the explanation of the physical phenomenon, the objective one used by the scientists.

The United Nations Intergovernmental Panel on Climate Change (IPCC), made up of scientists, defines the phenomenon as coming from any cause, human or natural.

This carries the implication that the world might cool as readily as it might warm up - an understandable assumption given the scientific evidence and the definitions we are given by the scientists.

Politicians are less precise with their rhetoric. Their policy statements tend to confound the scientific explanation.

The United Nations Framework Convention on Climate Change, the forum for the politicians, defines the phenomenon as attributed solely to human causes. The implication is that the world will warm and this warming is because of human actions.

However, during the first term of the George W Bush administration, politicians, commentators and, increasingly the media, dropped the term global warming in favor of climate change.

Why? Well that depends on how cynical you want to be. And how cynical depends on whether you think that the global climate change we are experiencing is due to human action or is just part of the normal scheme of events on a dynamic Earth.

The important thing is to realize that, be it objective fact, concept or fanciful idea, climate can get warmer and it can get colder.

It can get wetter or drier; stormier or calmer; more predictable or more uncertain. And all of these are aspects of climate change that usually do have many, often interacting causes.

The politicians and their spin-doctors were clever. They dropped global warming in favor of climate change when they realized that the consensus view of global warming held the anthropogenic label.

Better for them to widen the discourse, otherwise they would be forced by public pressure to try to do something about the warming, actions that would be costly and difficult to agree on. They dropped global warming for the more general and, they thought, less contentious term climate change.

CHAPTER 12

GETTING REAL ABOUT CLIMATE CHANGE

As some leaders in the field have suggested it's time to focus more on climate change adaptation, as the window for climate change prevention is arguably in the past.

Before beginning, though, it's useful to point out that, historically, unforeseen events have often pushed trends in directions that might have seemed unimaginable prior to their occurrence.

Scholars have documented individual and collective barriers, exploring social and psychological cognitive biases that create resistance to values. The mythical fixed pie bias

Over-discounting the future

Egocentrism

Positive illusions

Over-confidence

Pseudo-sacredness

The mythical fixed pie refers to a limited resources notion the authors regard as a fallacy, that if one party wins the other loses, instead of considering the possibility that both sides to satisfy their interests.

Negotiations reach an impasse because of a belief that each side is in perfect opposition to the other, and therefore tradeoffs are not thought to be possible.

Overly optimistic perceptions of oneself and the future, as compared to fact is a positive illusion that explains why companies promote as sustainable products of environmentally or socially questionable value or benefit.

People generally rate themselves higher on environmentally positive behaviors allowing them to maintain a more positive image of themselves.

Overconfidence in one's ability to estimate, and the disinclination to recognize and factor in uncertainties is another cognitive bias that leads to over-consumption and other environmentally destructive behaviors.

What is thought to be sacred is believed to be beyond negotiation or change, but not all that is considered sacred truly is, and what exists in this realm may be negotiable.

Here are other reasons the for-profit sector is a logical big player in climate change adaptation:

1. As technological innovators, companies best understand the economic and technical tradeoffs involved.

2. Companies must be involved in regulatory and policy decisions as government agencies do not have the knowledge or resources to develop the best solutions.

3. As social structures businesses, industries, and markets have accumulated power and resources to influence not only economic, but also social, environmental, and political conditions, and have been involved in developing solutions to problems in these realms.

4. Businesses can profit through creating innovations to satisfy societal preferences for products and services that resolve social and environmental problems.

Back to climate change adaptation, it appears that humankind is responding too slowly and ineffectually to reports of climate change documentation. Unfortunately, those who are the least responsible for carbon emissions are those who are most likely to suffer soonest due to climate change.

Vulnerable island populations in developing countries are seeing sea level rise and coastal flooding. The least developed areas of Africa and Asia likely to be more severely affected by drought and heat waves, leading to widespread worsened food scarcity problems.

Whether there is a direct link to Hurricane Sandy from climate change, that storm showed us in the United States that those most likely to suffer impacts are those who can least afford them.

Developed countries have used their wealth and power to resist agreements that might put them at a perceived comparative disadvantage, and this is unlikely to change.

Systems scientists tell us that the bigger the system, the slower the change, which is why it's so scary that the biggest global system of all, the biosphere, is changing pretty rapidly, even faster than earlier (and even recent) predictions suggested. The global human social system is also big, and change is likely be incremental and inconsistent.

As callous as it may seem to suggest it, it may be that the one influence that will create faster change is if those in positions of wealth, power, and influence suffer significant losses due to climate change related events.

National disasters like Hurricane Sandy, hitting Wall Street, one of the biggest economic engines in the world, maybe the kind of catalysts we need to stop dragging our feet.

CONCLUSION

Indeed the climate is changing; it has been doing so for thousands of years, and is likely to continue. Global warming happens, so does global cooling.

There is nothing alarming about it. Just as butterflies metamorphose and mountains change heights, climate also changes on different time scales.

Climate has changed partly because of man-made carbon dioxide. This I fully agree with. But more than the reasons, it is the consequences we have to look upon.

This change has not been, as many, say exponential. It is fair to call it 'normal'. I also do not think that the current climate policy will prevent it.

The global climate record which we use to make an estimate of this change and the presumable damage that will follow is contaminated.

The fact is - Satellites are showing less temperature than the surface thermometers, which we use in formulating these records, are.

The adjustments made, are based on mere estimation.

The Question about Climate Change is relative, i.e. before saying whether climate is changing or not, one has to specify a time interval, say for example 10 years, 100 years, or 1000 years. But more we go into history; the less valid is our assertion.

Nearly everyone agrees with the climate is warmer now than during the little ice age (some 200 years ago), but to say that it is warmer than what it was a thousand years back is incorrect. The data is all murky: it all depends on which set one accepts as truly representing global climate.

So, if there actually has been a change in the last decade or so, what might have caused it? Humans? Yes, but partly, as I said in the beginning. Solar activity is also one of the contributors to the changes we see. The astonishing part about this change is this - large part of this global climatic change has been positive. It has been helpful to mankind.

The warming so far has increased global vegetation cover, lengthened growing seasons, increased precipitation, and caused minimal ecological change.

Even the extreme weather events haven't increased much. So, to say that future climate change will be such that it would cause net harm than net good is incorrect.

It is really a shame to have to trout out statist arguments like - we'll be flooded, we'll be burnt due to heat... when pigs fly. There is no evidence to suggest that ecosystems will fail to adapt. Take for example malaria, which they say breaks out during high temperatures. You know, malaria retreated rapidly even as temperatures rose during the twentieth century.

And the worst outbreak of malaria in the 20th century happened in Siberia, before any significant human contribution to warming. Today it is prevalent in those countries which wouldn't run DDT based eradication programs.

There is one trend that is to be seriously looked upon: The Arctic ice. The melting of large volumes of ice in Polar Regions has more to do with black carbon (soot) than carbon dioxide.

Soot from dirty diesel engines and coal-fired power plants is now reckoned to be a far greater factor in climate change.

However, it is a short-lived pollutant that can be dealt with local rather than global action. Countries which waste billions in climate-change policies are beginning to realize it.

The 'climate policy' was economy-damaging to begin with but because Europe was relatively wealthy, they were able to live with it. But they too can no longer afford such canards.

Hence, the Question of whether we are too late on climate change is wrong. It presupposes climate change has actually taken place to such a damaging degree that it has become a mere Question of if we are late on it or not.

Any attempt to control climate is foolish. Or it is evil played by the state on citizens who're made to believe this idiocy. It is a good tool to destroy personal liberty. It is a scheme to justify additional taxes. It is a policy that hurts the poor who cannot afford increased prices in energy, and food.

BEST WISHES!

www.ingramcontent.com/pod-product-compliance
Lightning Source LLC
Chambersburg PA
CBHW070123230526
45472CB00004B/1395